SHOWING COWS AT THE FAIR

Gareth Stevens
PUBLISHING

By Jennifer Wendt

Please visit our website, www.garethstevens.com. For a free color catalog of all our high-quality books, call toll free 1-800-542-2595 or fax 1-877-542-2596.

Cataloging-in-Publishing Data

Names: Wendt, Jennifer.
Title: Showing cows at the fair / Jennifer Wendt.
Description: New York : Gareth Stevens Publishing, 2019. | Series: Blue ribbon animals | Includes glossary and index.
Identifiers: ISBN 9781538232866 (pbk.) | ISBN 9781538229286 (library bound) | ISBN 9781538232873 (6pack)
Subjects: LCSH: Cows--Showing--Juvenile literature. | Cows--Juvenile literature. | Livestock exhibitions--Juvenile literature.
Classification: LCC SF215.W46 2019 | DDC 636.2--dc23

First Edition

Published in 2019 by
Gareth Stevens Publishing
111 East 14th Street, Suite 349
New York, NY 10003

Copyright © 2019 Gareth Stevens Publishing

Designer: Katelyn E. Reynolds
Editor: Emily Mahoney

Photo credits: Cover, p. 1 (cow) TODD17/Shutterstock.com; cover, p. 1 (background photo) chainarong06/Shutterstock.com; cover, p. 1 (blue banner) Kmannn/Shutterstock.com; cover, pp. 1-24 (wood texture) Flas100/Shutterstock.com; pp. 2-24 (paper) Peter Kotoff/Shutterstock.com; p. 5 (Guernsey) critterbiz/Shutterstock.com; p. 5 (Brown Swiss) Fridi/Shutterstock.com; p. 5 (Ayrshire) R-Mac Photography/Shutterstock.com; p. 5 (Jersey) Lakeview Images/Shutterstock.com; p. 5 (Holstein) Marc Venema/Shutterstock.com; p. 7 © iStockphoto.com/Jacqueline Nix; p. 9 © iStockphoto.com/mikedabell; p. 11 Michael G McKinne/Shutterstock.com; p. 12 © iStockphoto.com/rusty13599; p. 13 John Patriquin/Portland Press Herald via Getty Images; p. 15 Jakkrit Orrasri/Shutterstock.com; pp. 17, 19, 21 © iStockphoto.com/BrandyTaylor; p. 20 © iStockphoto.com/Gannet77.

Printed in the United States of America

CPSIA compliance information: Batch #CW19GS: For further information contact Gareth Stevens, New York, New York at 1-800-542-2595.

CONTENTS

Words in the glossary appear in **bold** type the first time they are used in the text.

FUN AT THE FAIR!

There is so much to do at the fair! You can go on rides, watch shows, eat a snack, and even show animals. If you raised your cow to be a show animal, you might bring him or her to the fair to be judged!

Getting your cow ready to show at the fair can be a fun and exciting adventure. It takes a lot of work, but in the end, you and your cow might be blue-ribbon winners! Read on to learn how to take home first place!

There are many **breeds** of cow you can show at the fair. You can show both male and female cows.

GUERNSEY

BROWN SWISS

AYRSHIRE

JERSEY

HOLSTEIN

WEARING A HALTER

Before showing your cow at the fair, you'll need to get her used to wearing a **halter.** Once you get the halter on your cow, leave it on for a few days so she gets used to it.

It's important that your cow has a halter that fits the right way. This will help you control her movements in the show ring. Your cow's halter needs to be comfortable so she'll be relaxed and listen to you.

TAKE THE PRIZE!

YOU CAN ALSO RUB YOUR COW'S HEAD, FACE, AND EARS SO SHE GETS USED TO BEING TOUCHED.

You'll need a halter and **lead** to show your cow at the fair.

WALKING NICELY

Once your cow is used to wearing a halter, it's time to take her for a walk. Walking your cow several times a week will help her learn she can't play tug-of-war. She should walk next to you with her head held high.

The fair can be a very noisy place. Think of some of the noises you might hear and try to get your cow used to them. This will keep your cow from getting scared in the show-ring.

TAKE THE PRIZE!

IT CAN TAKE UP TO 3 MONTHS FOR YOUR COW TO GET USED TO WEARING A HALTER AND LEAD, SO BE PATIENT!

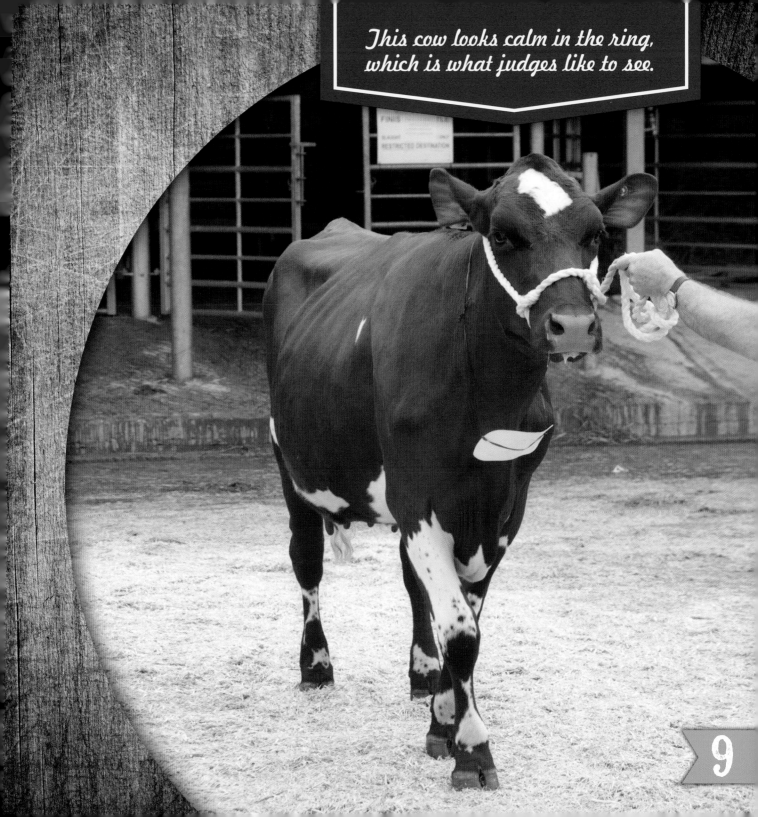

This cow looks calm in the ring, which is what judges like to see.

9

BATH TIME!

Washing your cow as you train her will help get her ready for the day of the show. This way, she won't be nervous when you **groom** her at the fair, and her coat and skin will stay healthy.

Using cool water will keep your cow's hair shiny. Use a gentle shampoo and make sure to rinse well. Brush her all over to remove any dead hair and pay special attention to her head, tail, and all of her white areas.

TAKE THE PRIZE!

A COW GROWS A NEW COAT OF HAIR EVERY 90 DAYS. YOU CAN COUNT BACK THE NUMBER OF DAYS FROM THE FAIR TO MAKE SURE ALL THE OLD HAIRS ARE BRUSHED OUT.

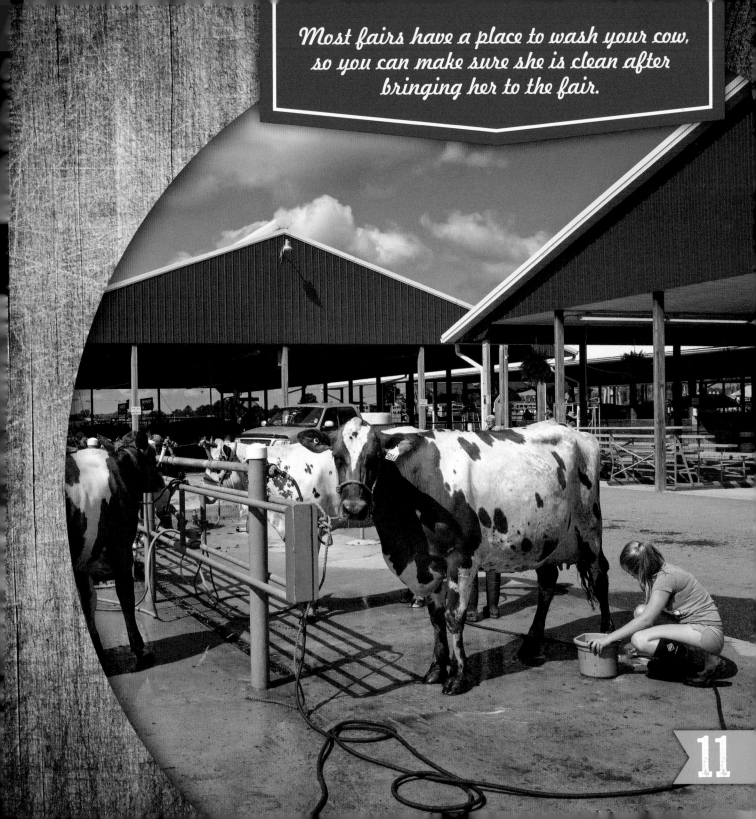

Most fairs have a place to wash your cow, so you can make sure she is clean after bringing her to the fair.

GROOMING YOUR COW

Grooming can be a bonding time for you and your cow. Just like you, your cow needs her ears and eyes cleaned, her hair **trimmed,** and her feet washed. She may also need her hooves clipped. Don't forget to fluff her tail!

Your grooming tools can be stored in a show box that you'll keep by your cow at the fair. A show box is like a big suitcase for your cow. You can pack everything she'll need at the fair into the box.

Some supplies you may want to keep in your show box include: a bucket, a brush, shampoo, an extra lead and halter, towels, and fresh food.

FEEDING YOUR COW

Feeding your cow the right amount of food is important. Most cows eat a mixture of corn and hay. You can ask your **veterinarian** about what kind of food is best for your cow.

Be sure to take the same food that you feed your cow at home to the fair so she is comfortable. Always make sure your cow has plenty of fresh water and lots of clean straw. Remember to clean up your cow's **manure** and change her **bedding** as well.

TAKE THE PRIZE!

A COW'S STOMACH ACTUALLY HAS FOUR PARTS! THE FOODS COWS EAT TAKE A LONG TIME TO **DIGEST**, SO EACH PART HAS A SPECIAL JOB TO DO.

Cows don't have top front teeth! Their mouths are made for eating grass.

15

SHOWMANSHIP

Showmanship is about how you present your cow to the judges. It includes how you handle your cow, how the two of you work together as a team, and how you both look and act.

Your cow needs to be clean for the fair, and so do you! Wear clean clothes, wash your hands, and brush your hair. You may also need to wear a uniform if you're part of a group that shows animals at the fair.

TAKE THE PRIZE!

YOUR COW'S AREA SHOWS HOW WELL YOU CARE FOR HER, SO YOU'LL WANT TO KEEP IT CLEAN. YOU MAY ALSO WANT TO BRING A POSTER WITH HER NAME ON IT AND SOME FACTS ABOUT HER!

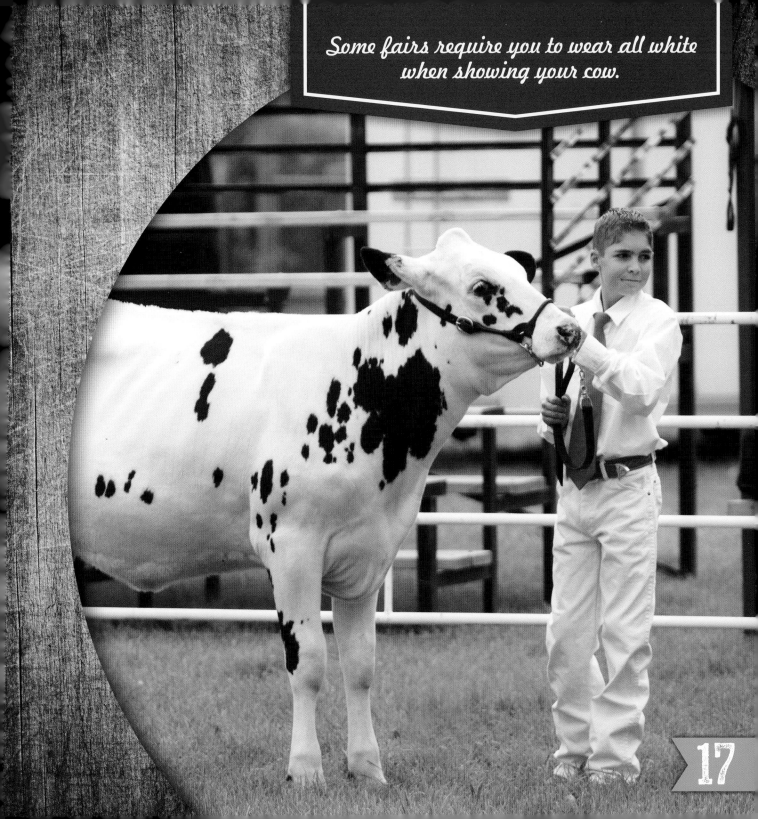

Some fairs require you to wear all white when showing your cow.

17

SHOW-RING READY

It's show time! Bring your cow's **vaccination** records and your fair **registration papers** when you check in. Settle your cow in her pen and get ready to have fun showing your cow!

When it's your turn to show, walk your cow to the judge **confidently.** Set your cow up to look her best. Look the judge in the eye when he or she talks to you, and be sure to answer questions clearly.

TAKE THE PRIZE!

SOME QUESTIONS A JUDGE COULD ASK ARE: WHAT KIND OF COW DO YOU HAVE? WHAT DO YOU FEED HER? HOW OLD IS YOUR COW? HOW LONG HAVE YOU HAD HER?

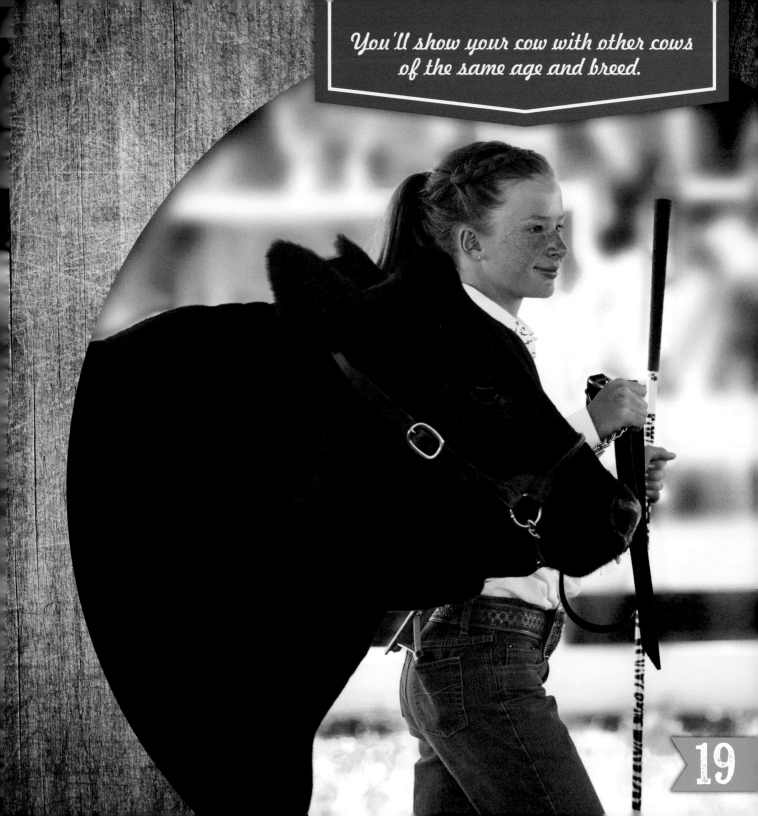

FINISHING UP

When the fair ends, hopefully you and your cow will go home with a ribbon! Not everyone wins a ribbon, but the fair is fun and you'll learn a lot about your cow. It's also a great place to make new friends!

Be sure to clean your area and thank everyone who helped you get ready for the fair. Double check to be sure you put all your supplies back in your show box. Say good-bye until next year!

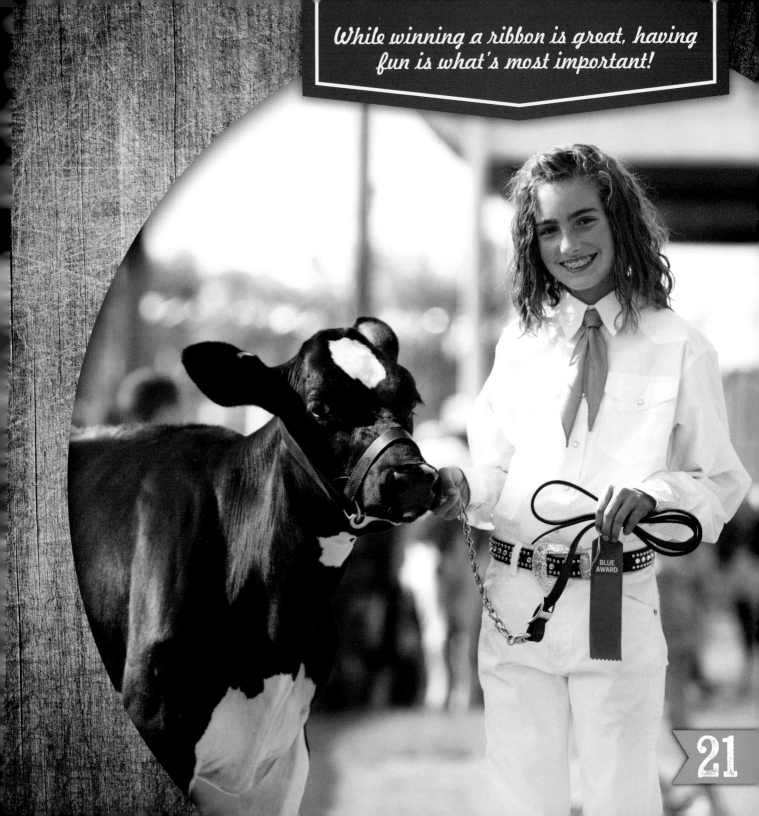

While winning a ribbon is great, having fun is what's most important!

GLOSSARY

bedding: matter used for an animal's bed, such as straw, newspaper, or wood shavings

breed: a group of animals that share features different from other groups of the kind

confident: having a feeling or belief that you can do something well

digest: to break down food inside the body so that the body can use it

groom: to clean

halter: a set of straps placed around an animal's head so that the animal can be lead

lead: a long, thin piece of rope that is used for holding an animal

manure: a cow's waste

registration papers: an official record of information

trim: to clip or cut down in size

vaccination: a shot that keeps a person or animal from getting a certain sickness

veterinarian: a doctor who is trained to treat animals

For More Information

BOOKS

Gibbs, Maddie. *Cows.* New York, NY: PowerKids Press, 2015.

Schuetz, Kari. *Cows.* Minneapolis, MN: Bellwether Media, 2014.

Silverman, Buffy. *Meet a Baby Cow.* Minneapolis, MN: Lerner Publications, 2017.

WEBSITES

4-H
4-h.org
4-H gives children a chance to learn new skills through hands-on projects.

Holstein Foundation
www.holsteinfoundation.org
Information on youth programs, leadership programs, and showing animals at the fair can be found on this site.

National FFA Organization
www.ffa.org
Future Farmers of America is an education-based organization for students interested in farming.

INDEX